Robin Kerrod

Series consultant:
Professor Eric Laithwaite

Educational consultant:
Dr Helen Rapson

Subject consultant:
Len Treharne

How to use this book

There are four books in the **Science Alive** series: **Moving Things, Changing Things, Living Things,** and **All Around.** They will introduce you to science in the world around you.

To find information, first look at the contents page opposite. Read the chapter list. It tells you what each page is about. You can then find the page with the information you need.

The colors on the contents page will help you to find your way through the book. Each main topic in the book is shown by a colored stripe, that matches the color around the edges of the pages about that topic.

On each pair of pages inside the book, you will see this in the top right hand corner:

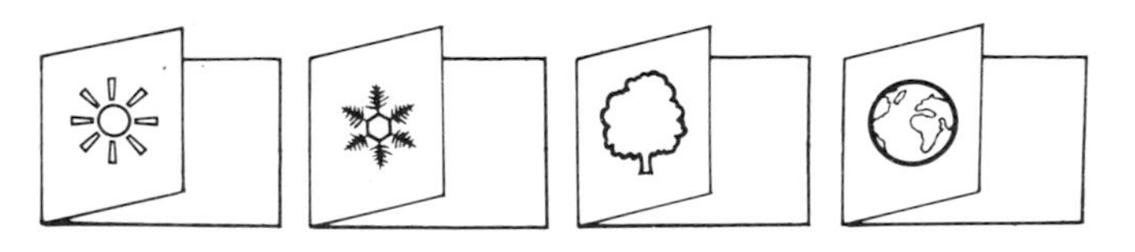

This shows you where to find out more about the subjects covered in those two pages. The signs tell you which of the four Science Alive books and which pages to look at.

 is the symbol for Moving Things,

for Changing Things,

for Living Things,

for All Around.

You can see the symbol on the cover of each book.

For example, pages 32–33 of this book are about building and have these signs:

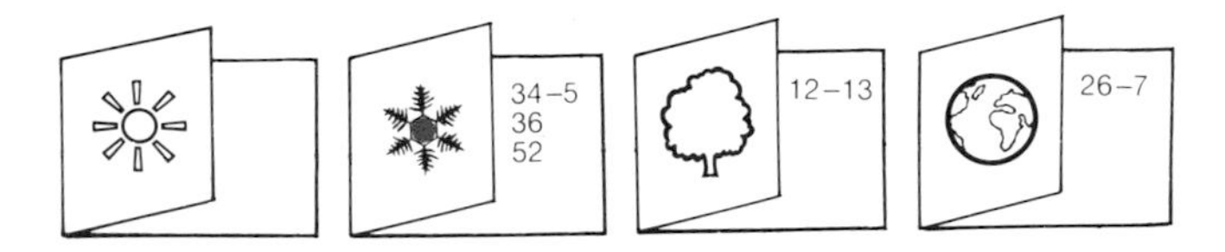

These mean: to find out more about building, look at pages 34–35, 36 and 52 of this book, pages 12–13 of Living Things, and pages 26–27 of All Around.

On pages 37–52 of this book there are some simple activities and experiments you can try. There is a separate contents list of those on page 37, and they will help you to understand and find out more about the information in the rest of this book.

If you want to know about one particular thing, look it up in the index on page 54. For example, if you want to know about crystals, the index tells you that there is something about them on page 9. The index also lists the pictures in the book.

When you read this book, you will find some unusual words. The first time each one is used it is written in **dark** letters. The glossary on page 53 explains what these words mean.

CONTENTS

Things around us

How many different things can you see around you? Do you know what they are made of? This book is made of paper, and the words are printed in ink. We make the paper from wood, which comes from trees. We make the ink from oil.

Many of the other things around you came from very different materials. Think of your radio. The outside is probably made of **plastic**, and that plastic comes from oil. The inside of the radio is full of copper wires. Copper is a metal which we get from **minerals**, and we dig minerals out of the ground.

Metal and wood feel hard and firm because they are both solid things. If you look at some fizzy soda, you will see two other kinds of things. The soda is a liquid, and the bubbles in the soda come from a gas. Everything in the world is one of these three things—a solid, a liquid, or a gas.

Water can easily change from a liquid to a solid or a gas. Water from the tap is a liquid, but if you freeze it, it changes to ice, which is a solid. If you heat it, it boils and makes bubbles of gas.

The kitchen is full of different kinds of things. They are all made from different materials. The three main materials are wood, metals, and plastics. What other materials can you find in the picture?

Solids

Solid things hardly ever change their shape or size. You can test this for yourself. Put two solid things, a piece of wood and a pebble, on a table. Measure them and remember what shape they are. Look at them again a few days later. Have they changed their size or shape? Will they change unless you cut them or smash them? This is one way in which solid things are different from liquids and gases.

Most of the things we find around us are solids. One solid you are sure to have in your home is sugar. If you look at sugar closely with a magnifying glass, you will see that it is made up of thousands of tiny colorless pieces. These are the **crystals** of sugar.

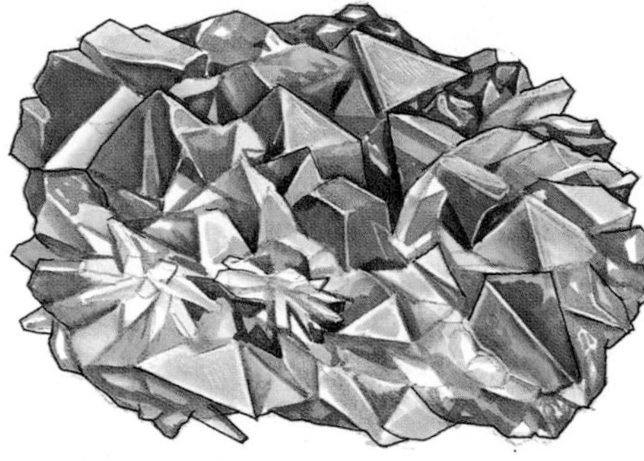

pyrite

These are all the kinds of crystal you might see.

quartz

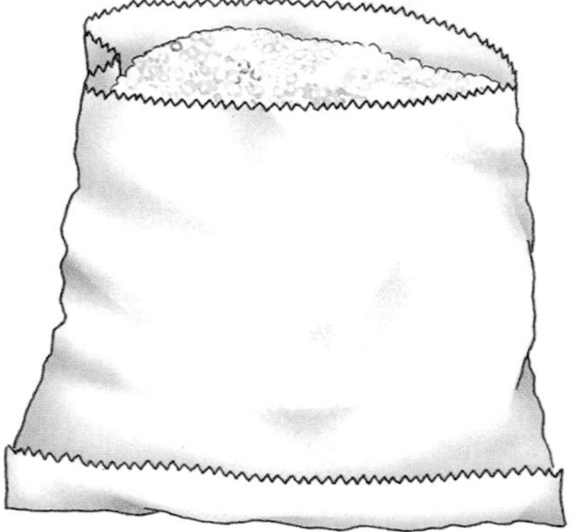

Italian ice

This cliff is made of a rock called granite. Granite is a solid. It is hard and will not change for a long time.

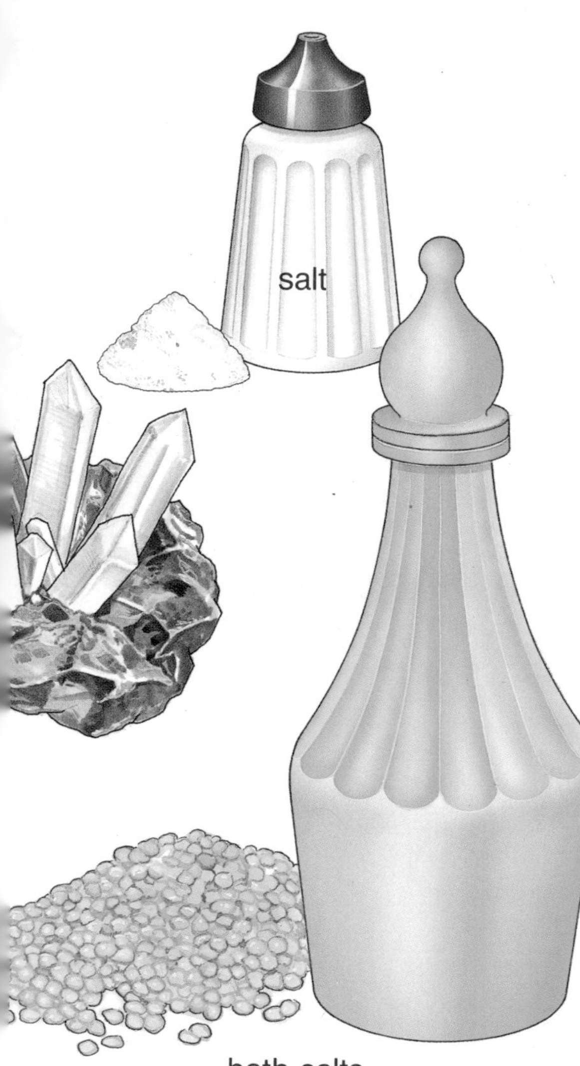

Many other solid things make crystals, too. You can see shiny crystals in many rocks. The different minerals in the rock make crystals in different colors and in different shapes. Some minerals make crystals shaped like tiny blocks. Others make long pointed crystals, rather like sharpened pencils.

In most rocks, the different shaped crystals are pressed tightly together. In some rocks, there are holes and here the crystals have space to grow into the most beautiful shapes.

Ice can also make beautiful crystals. You may see them on your window on a frosty morning, making feathery patterns as they grow along the glass.

These valuable crystals, or gems, have been cut so that they sparkle.

Gases

There are gases all around us. They are the gases that air is made of. You can't see the air, you can't touch it and you can't taste it. But you can feel it and hear it when it moves. Moving air is called wind. You can also feel the air pushing against you when you cycle or run fast.

Air is one of the most important things on Earth. We must breathe air to live. Without air, we would die. Underwater and out in space there is no air. We can only go there if we take air with us.

People who dive under the water take air with them in metal bottles. We can squeeze a lot of air into the small bottles. This is called **compressed air**. We can squeeze all gases in the same way, but we can't do this to liquids or solids.

Deep underwater, a diver goes exploring. She breathes air through a tube from the air bottles on her back.

26–7
6–7
40
20–21
8–9
38–9

You squeeze, or compress, air into a tire when you blow it up.

Many gases have no smell, but some are very smelly indeed. Have you ever cracked open a bad egg? If so, you have smelled one of the nastiest gases of all!

Another smelly gas is the one we use for cooking, in gas stoves. We call it **natural gas** because we don't make it ourselves. We find it in nature, trapped in rocks underground.

If we squash natural gas, like air, something interesting happens. The gas turns to liquid, so we can put it in metal bottles and carry it around. We use it for cooking, heating, and lighting when we go camping, in motor homes, or on boats.

When you turn the tap. some of the liquid gas in this bottle turns back to gas and comes out.

Liquids

How many liquids can you think of in and around your home? It's easy to guess the most common one. It's water, of course. Many of the other liquids you will think of are also made mainly of water, such as juices, soft drinks, milk, tea, and coffee. But how many other different liquids can you think of? Here are some clues. What do we use in washing dishes, washing our hair, for cooking, in cars, and in thermometers?

Liquids can flow from one place to another. Think of what happens when you open a bottle of milk and tip the open end over a glass. The milk pours out of the bottle and into the glass.

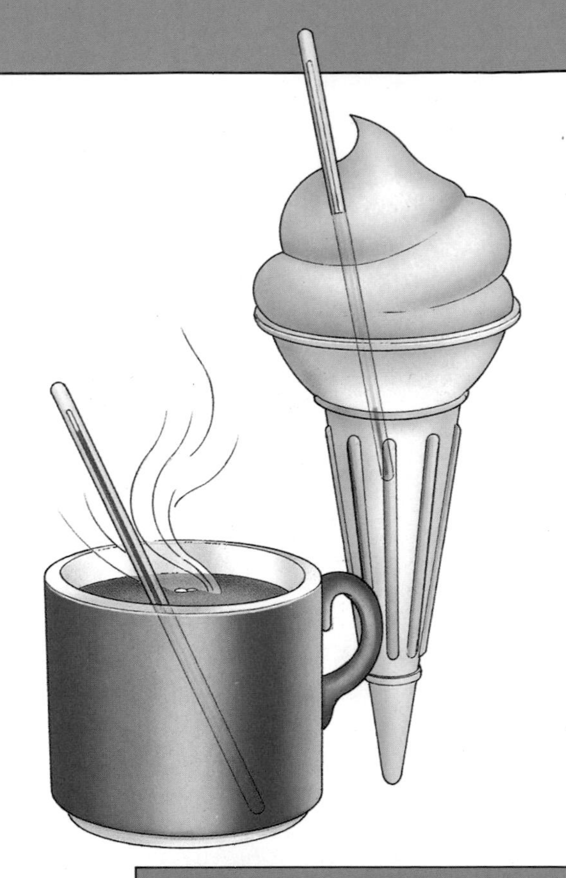

The liquid in the thermometer goes up in hot things and down in cold things.

brake pedal

piston pushes on liquid

liquid pushes brake pads against disk

When you press the brake pedal of a car, a liquid inside pushes the brakes against the wheels and stops them.

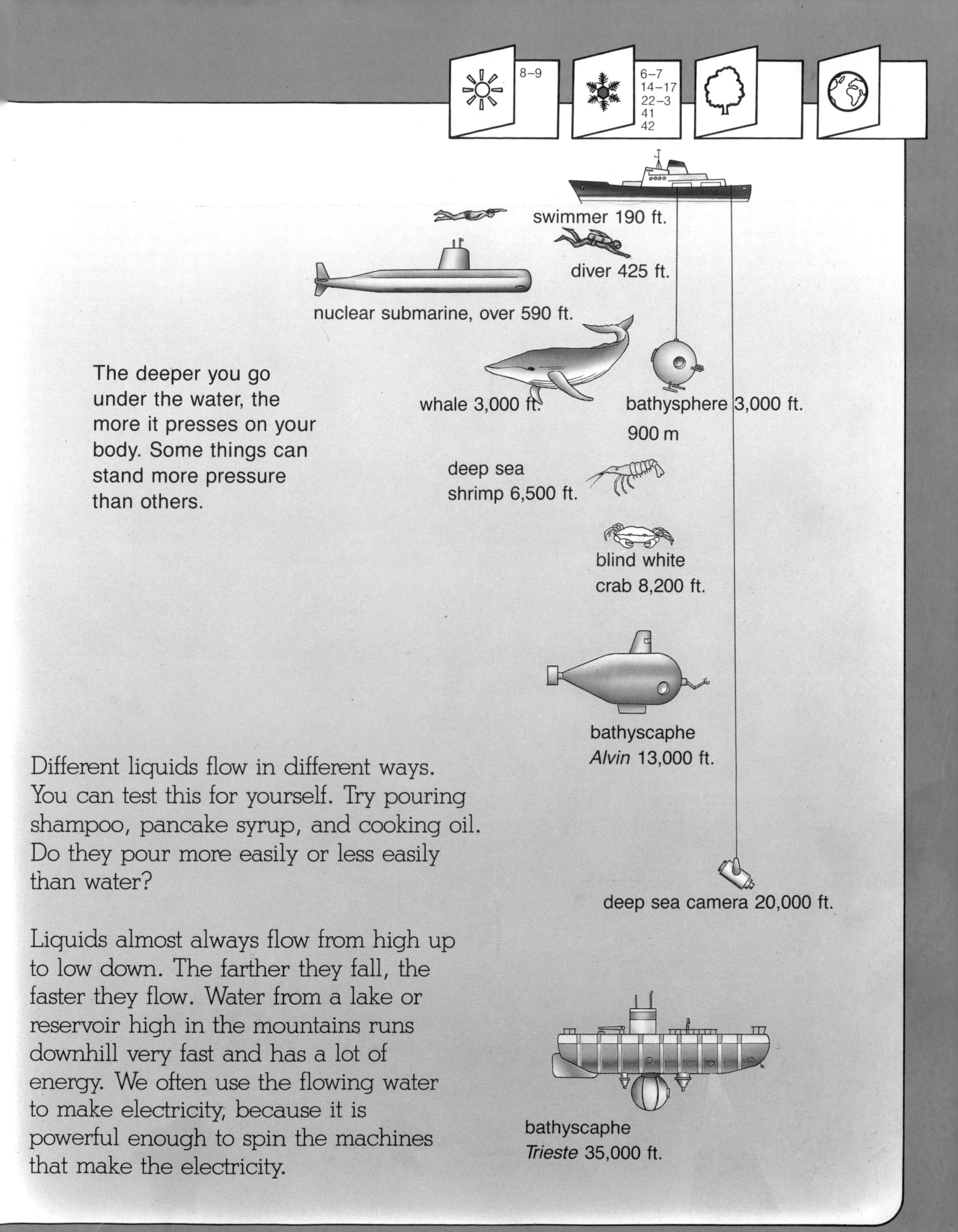

The deeper you go under the water, the more it presses on your body. Some things can stand more pressure than others.

Different liquids flow in different ways. You can test this for yourself. Try pouring shampoo, pancake syrup, and cooking oil. Do they pour more easily or less easily than water?

Liquids almost always flow from high up to low down. The farther they fall, the faster they flow. Water from a lake or reservoir high in the mountains runs downhill very fast and has a lot of energy. We often use the flowing water to make electricity, because it is powerful enough to spin the machines that make the electricity.

The skin on liquids

Take a glass and fill it with water. When it is full right up to the top, stand it on a kitchen table and try adding more water to it, a drop at a time. Do this carefully and without shaking the glass. You will find that you can keep adding the drops until the water rises above the rim of the glass.

Look at the top of the water and you will see a kind of skin stretching across it, which stops the water from spilling out. All liquids have a kind of skin. We call this **surface tension**.

Look closely at the top of a pond and you will usually see signs of the skin there, too. You may see little flies walking around on top of the water and not sinking. You may also see insects called pond skaters, gliding across the water.

A pond skater has tiny hairs on its feet. They stop it from piercing the water's surface.

You can also see the surface tension working if you try to make a sieve hold water. Pour only a little water into the sieve very gently. The water will fill the holes in the sieve and bulge through them, but should stay there if you have been careful. Now try putting water in the sieve again, but mix a few drops of liquid detergent with the water first. What happens this time?

When you go camping, you have to be careful not to break the surface tension of raindrops on the outside of the tent. If you touch the tent inside, the surface tension will break and the water will soak through.

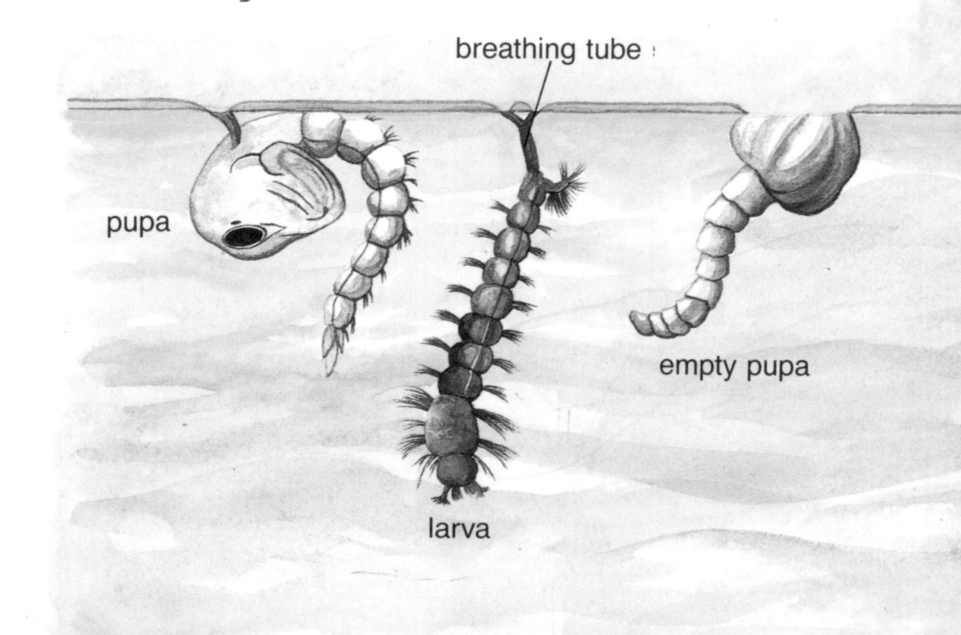

The surface skin on water is very important to mosquitoes. They usually lay their eggs on the water. When the young larvae hatch, they hang onto the surface skin with their breathing tubes sticking out.

Water

There is water almost everywhere on Earth. There is water in the sea, in rivers, in lakes, in the soil, and deep under the ground. There is frozen water in the great icefields at the North and South Poles. There is water gas, which we cannot see, in the air around us. It is called water vapor. Even the clouds are made of drops of water or specks of ice.

There is also water inside us. Over half of what we weigh is water. If you weigh about 65 pounds, you have over 40 pounds of water inside you!

People, plants, and animals all need water to live. People and animals drink it, plants need it to make their food.

We use a lot of water every day. We drink it by itself or in juice or soft drinks, we wash in it. We use it to flush the toilet.

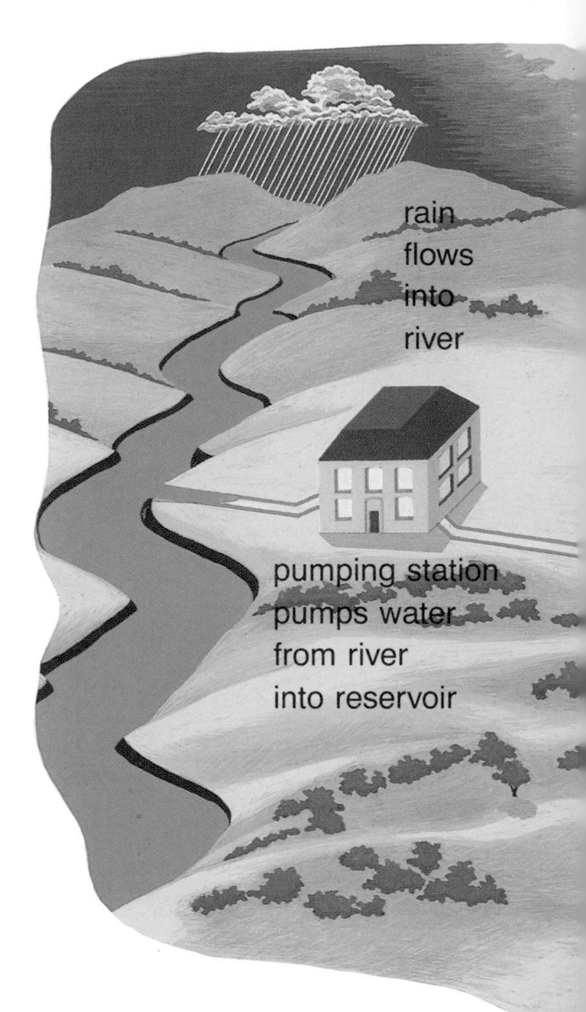

This is what the Earth looks like from space. It is blue because that is the color of the sea, and the sea covers most of the Earth.

26–7
36
47
6–7
12–15
18–19
22–3
42–3
20–21
6–7
12–15
20–21
22–3
26–7

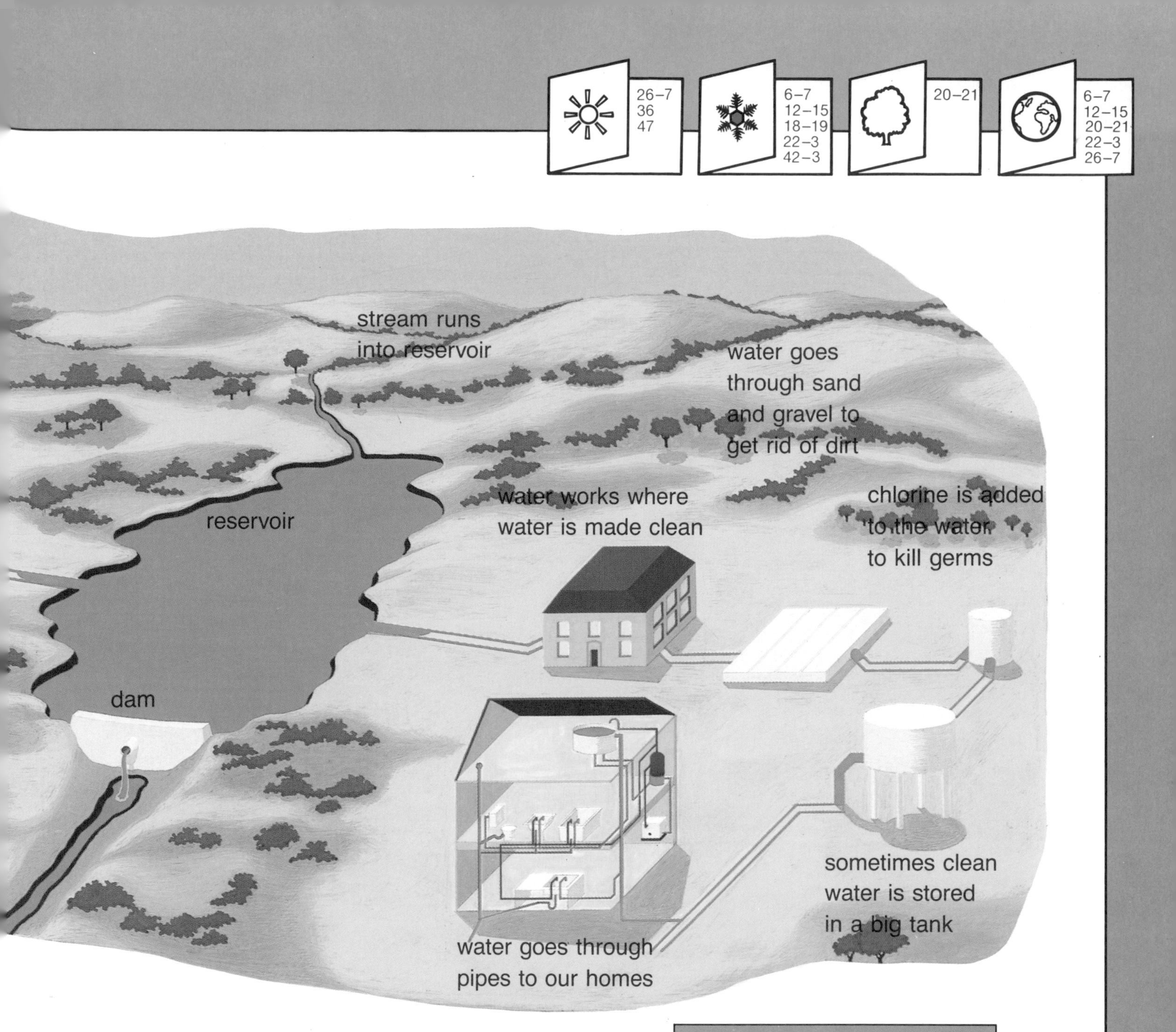

How much water do you use every day for washing and drinking? Try to find out. You may need a grown-up to help you. Find a bucket with a scale marked in quarts on the inside. Use it to fill up your wash basin and kitchen sink so that you can measure how much water they hold. Use a measuring cup to measure how much your glass and cup hold. Keep a note of how many times you use them all in one day. Can you figure out how much water you use in a year?

This is the journey tap water makes to most homes. It often comes from rivers. We collect it in **reservoirs** and clean it at the water treatment works. Sometimes we store it in towers, before it goes through pipes to wherever we want.

Floating and sinking

Find a piece of wood and a stone of about the same size, and put them in a bowl of water. What happens? The wood floats and the stone sinks. Why? You may say it's because the wood is light and the stone is heavy. You are nearly right: wood floats because it is light for its size; stone sinks because it is heavy for its size. We say that stone is more dense than wood. But that's not all.

The weight of the wood pushes it down against the water, but the water pushes back. The pushing up of the water balances the pushing down of the wood, and stops the wood from sinking. The stone is heavier so it can push harder and the water cannot stop it from sinking.

These lines show how far a ship can be loaded in different types of water. TF stands for tropical fresh water; F for fresh water; T for tropical sea water; S for summer in sea water; W for winter; WNA for winter North Atlantic.

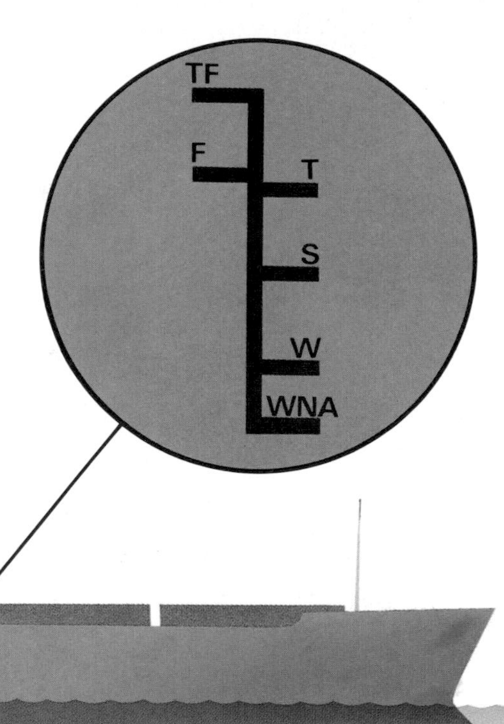

Ships are heavy but their shape helps them float. They are shaped so that the push of the water balances the push of their weight.

Lines on their sides show how far ships sink. They sink lower in fresh water than in salty sea water, and not as far in cold as in warm seas.

28–9
43
44–5

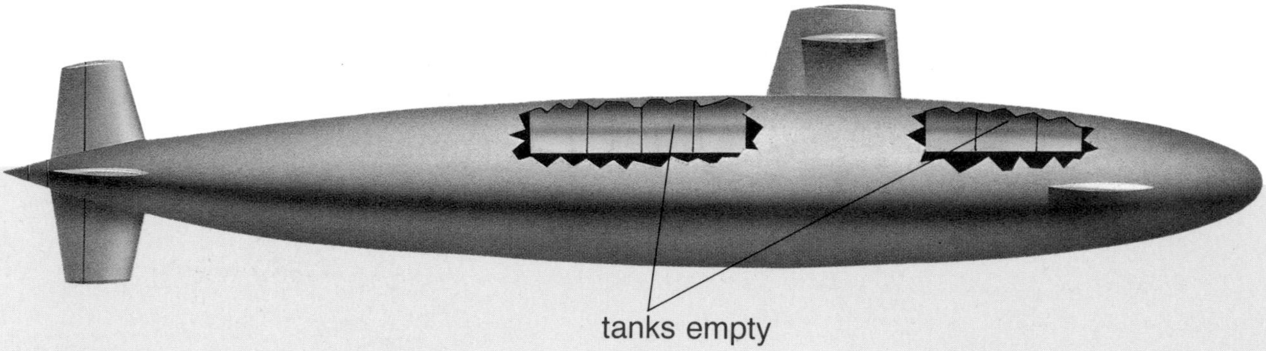

tanks empty

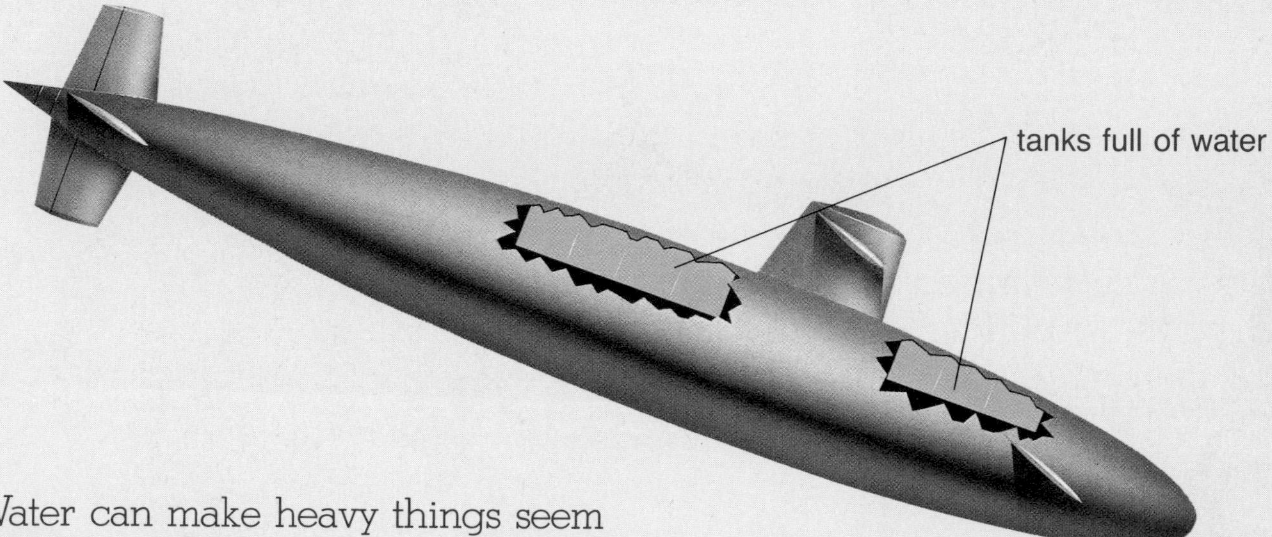

tanks full of water

Water can make heavy things seem lighter. Find a heavy stone, tie it on a string and lower it into a bucket of water. The stone will feel a lot lighter, because the water is pushing up against it.

Does an egg float in water? Try floating one in a jar of water, and you will find that it sinks to the bottom. But you can easily make it float. Add several big spoonfuls of salt to the water and stir it well. Put in the egg again, and this time it should float. Add more salt if it doesn't. The egg floats because salty water has a much stronger upward push than plain water has.

Submarines travel under the water. They sink by letting water flood into their tanks. They come up again by pumping the water out.

Mixtures

A lot of the things we see every day are a mixture of many things. Soil is a mixture of sand, rock chips, wood, and rotting plants. Most rocks are made up of a mixture of different minerals. In granite, a very hard rock, you can easily see the minerals as different colored crystals.

Look carefully at some soil and you will probably see the different parts of the mixture. But they are well mixed because they are such tiny pieces. You could easily mix sand and sugar, too, because they are made of small crystals, but you would still be able to see the different colors of the sand and sugar. If you mixed salt and sugar, it would be harder to spot the different parts of the mixture, but you would still be able to taste both salt and the sweet sugar.

Oil and water don't mix very well. Oil floats on top of the water as an oil slick. When it leaks into rivers or the sea, the oil can cover the feathers of water birds and kill them. It can be washed ashore where it pollutes the beaches so that they are unpleasant to visit.

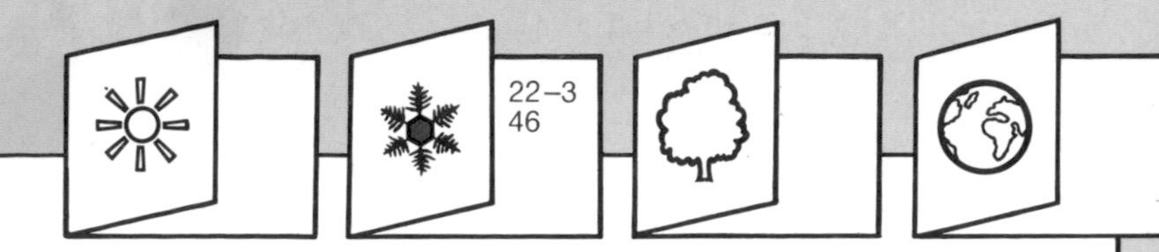

Some liquids don't mix well at all. Oil floats on top of water. Milk is a mixture of tiny fat droplets in water. When you leave it standing, some of the fat rises to the top, as cream.

But other liquids mix very well. Orange juice mixes with water so that it has only an orange taste. Melting things to make them liquid makes it easier to mix them. This is how metals are mixed to make new metals, called **alloys**, which are more useful to us than the separate metals. Alloys may be stronger, or less likely to **rust**, or harder. You can't see the different metals in an alloy.

airplane built with duralumin, an alloy of aluminum and copper

bronze statue (an alloy of copper and tin)

THINGS MADE FROM ALLOYS

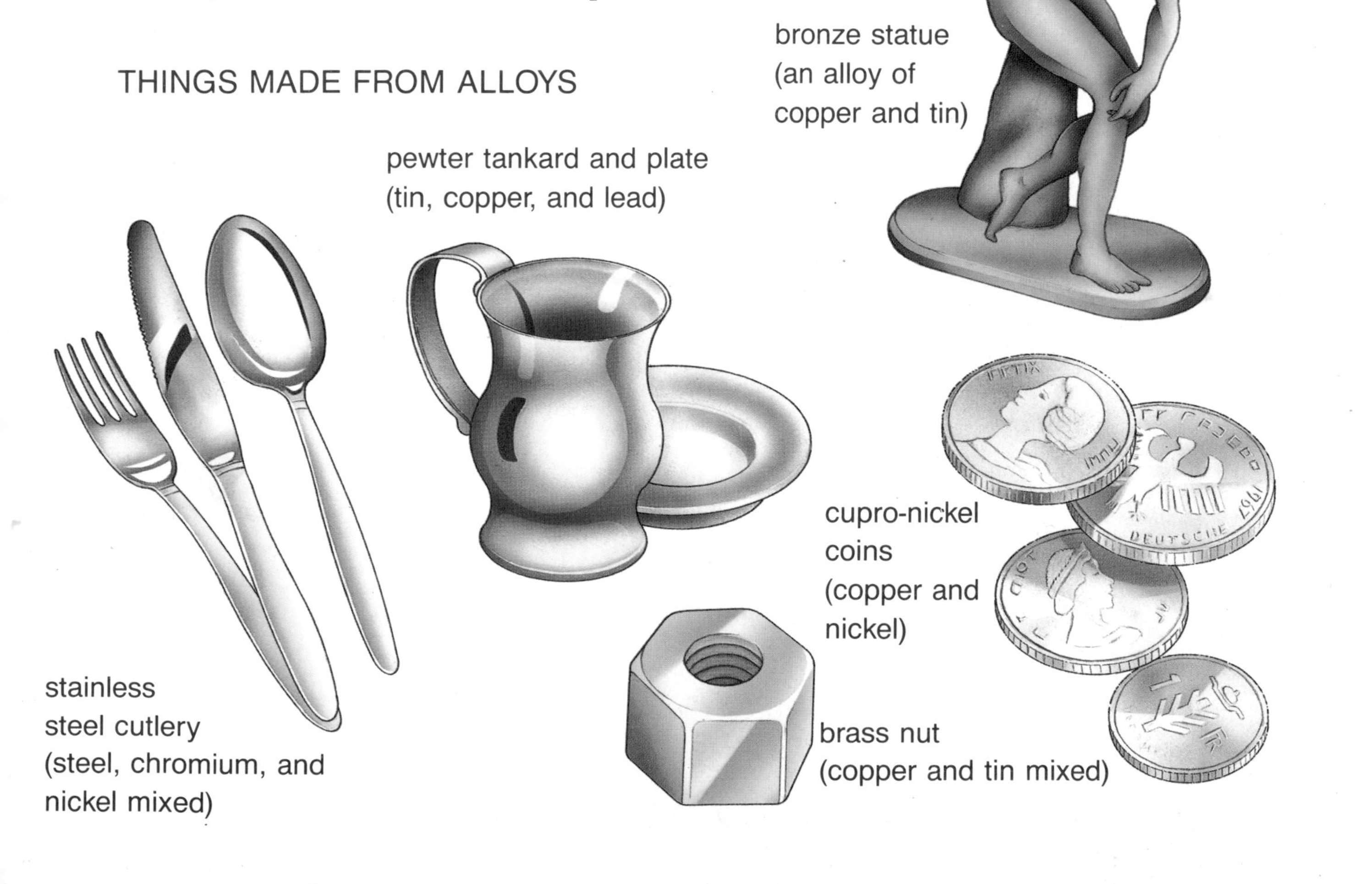

pewter tankard and plate (tin, copper, and lead)

stainless steel cutlery (steel, chromium, and nickel mixed)

cupro-nickel coins (copper and nickel)

brass nut (copper and tin mixed)

Solutions

Pour some water into a glass and put in a big spoonful of sand. Stir the water and the sand with a spoon. What happens? The sand just sinks to the bottom of the glass. Now pour some water into another glass, and add a big spoonful of salt. Stir it and the salt will disappear. Even if you look with a magnifying glass, you will not see any salt. But if you dip your finger in and taste the water, it will taste salty. The salt is still there somewhere, but you cannot see it. It has mixed into the water, or dissolved, to make a **solution**.

Water can **dissolve** other things, such as sugar, lemonade crystals, bath salts, and baking powder. It cannot dissolve sand or grease. To dissolve grease, you need household cleaning fluid. This is often called a **dry cleaner** because there is no water in it.

In many underground caves there are great stone "icicles" hanging from the roof and pillars rising from the floor. They are made when dissolved limestone is left behind by dripping water.

If you leave pools of sea water in the sun for long, the water dries up. Only the salt from the water is left.

46
47
18–19
20–21

There are things dissolved in water everywhere. For example, our blood is mostly water. Many things are dissolved in it, and the blood carries them to the parts of the body that need them. The sea has salt dissolved in it, and many rivers have minerals in them. The water dissolves the minerals in the rocks when it flows over them.

If the water which comes to our homes has a lot of things dissolved in it, we call this **hard water**. **Salt water** does not have many things dissolved in it. You can tell if your tap water is hard by looking inside the kettle. If it is hard, the inside will be coated with a kind of white foam. This is the stuff that has come from the water when the kettle boiled.

Changes

Take some new shiny iron nails and put them on an old saucer outside. Look at them every few days. If the weather is very damp, you will see that the nails soon turn brown on the outside. If you leave them long enough they will have small holes in them. The iron is rusting away as it joins up with part of the air to form rust.

Gases and moisture in the air are changing the surface of this statue.

Rusting is an example of change. There are changes going on around and inside us all the time. Cooking food makes it change. It tastes better, or it is easier to eat. If you make a loaf of bread, you wouldn't want to eat the flour, salt, and yeast on their own, but when they are mixed with water and baked, the mixture tastes good. When we eat food, we change it again. Our bodies break it down and use it for energy and to grow.

Acids in the air have started to eat away this stone.

When wood and coal burn, they are changing. They give off heat and light, and leave ashes behind. Burning things can give more than just heat and light. They can give off gases, or fumes, and soot into the air. This can form a dirty black coating on buildings. But sometimes the fumes join with rain to make an acid, like lemon juice. Acid rain falling on plants and into lakes can kill the wildlife. Acid rain can also damage buildings. It dissolves away the stone, or **corrodes** it. All these are chemical changes.

We can see many other changes like those. Salt splashing up from roads in winter corrodes cars. People use chemical changes, for example, to clean their false teeth by dissolving away any food, and to change the color of their hair by bleaching it.

Forest fires burn fiercely and spread very quickly. They can kill many plants and animals. But not long after this fire, plants began to grow again.

Different materials

Have you ever wondered why things are made the way they are? For example, why are most pans made from aluminum? Why aren't they made from **plastic**?

The reason is that aluminum and **steel** do some things better than plastic or lead. Aluminum is good for making pans because it lets heat pass through easily and cooks the food quickly. Plastic would melt or burn when it was heated.

Thousands of other things are made from steel, such as needles, cars, railroad tracks, and bridges. But steel is quite heavy, so it is not good for making airplanes, which must be really light. Instead, we make them of aluminum, which is very light.

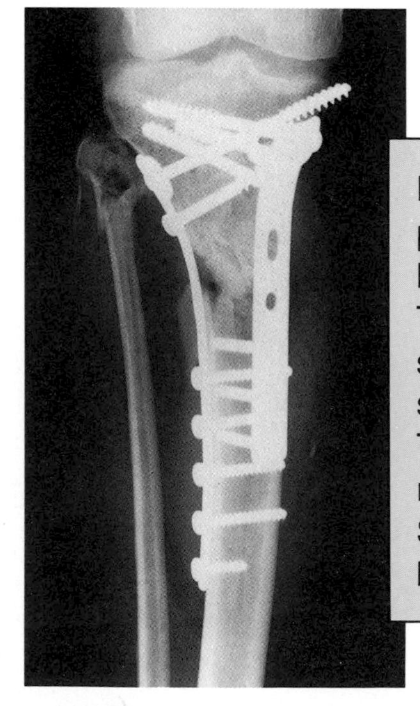

Doctors can put steel pins through a broken bone to keep it together. They must use stainless steel, because ordinary steel harms the body. These are X-rays of motorcyclist Barry Sheene's legs after a bad accident.

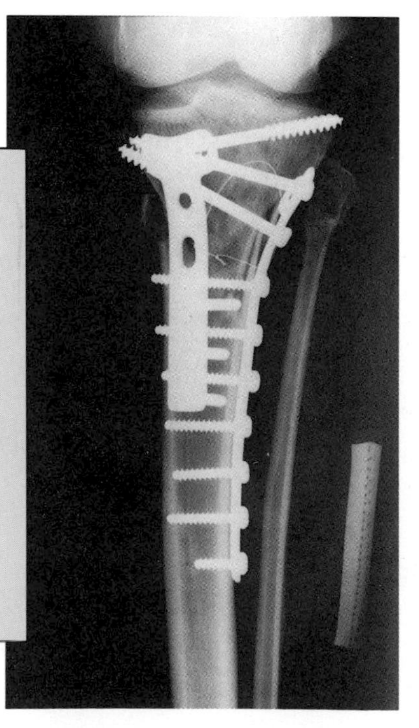

Pole vaulters use poles made of plastic because it can bend so easily. Fibers run through the plastic to strengthen it.

Aluminum is also better for airplanes because it doesn't rust, like steel. Rust is a brown powder. If ordinary steel gets damp, it slowly turns to rust and becomes weak. But we can make a special steel that doesn't rust, called stainless steel. We make it by mixing two other metals, chromium and nickel, with the steel, to make an alloy.

We use many alloys in our everyday life. We use **bronze**, an alloy of copper and tin, for coins because it does not rust or get damaged easily.

Plastics are almost as important as metals. How many plastic things can you see around you? We use plastic a lot because we can make things from it so quickly and cheaply.

bulldog clip
strip of paper being tested
yogurt cup holds weights
ruler
plastic tube
marble
sample being tested

Here are two ways in which you can test how strong and how hard things are. To test for strength, hold the material between two clips and then hang some weights from the bottom clip. You can test the hardness by dropping a marble and seeing how high it bounces.

Where do things come from?

Do you know that aspirins, nylon stockings, and car grease have something in common? They are all made from the same thing. It is an oil, called crude oil or **petroleum**. Petroleum comes from under the ground. We drill holes to reach it, and it bubbles up.

When it comes out of the ground, petroleum is just a **raw material** and we have to do things to it to make it ready to use. We call this **refining**, and we do it in an oil refinery. There, we make the oil into many different liquids and gases.

We use some of the liquids, such as gasoline, to make machines work. We turn the other liquids and gases into hundreds of different things.

Petroleum is one of our most useful raw materials. Many other raw materials come from the ground, too. Metals come from the minerals in rocks. We get them by heating the minerals in a huge, fiery oven, or furnace.

Clay also comes from the ground. It is like a sticky mud which we can shape and bake to make pottery such as cups, bowls, and tiles.

Wood is another useful raw material. We can use it by itself for building, or we can turn it into paper, plastic, or explosives.

Crude oil is a mixture of hundreds of things. Here are some of the goods made from crude oil.

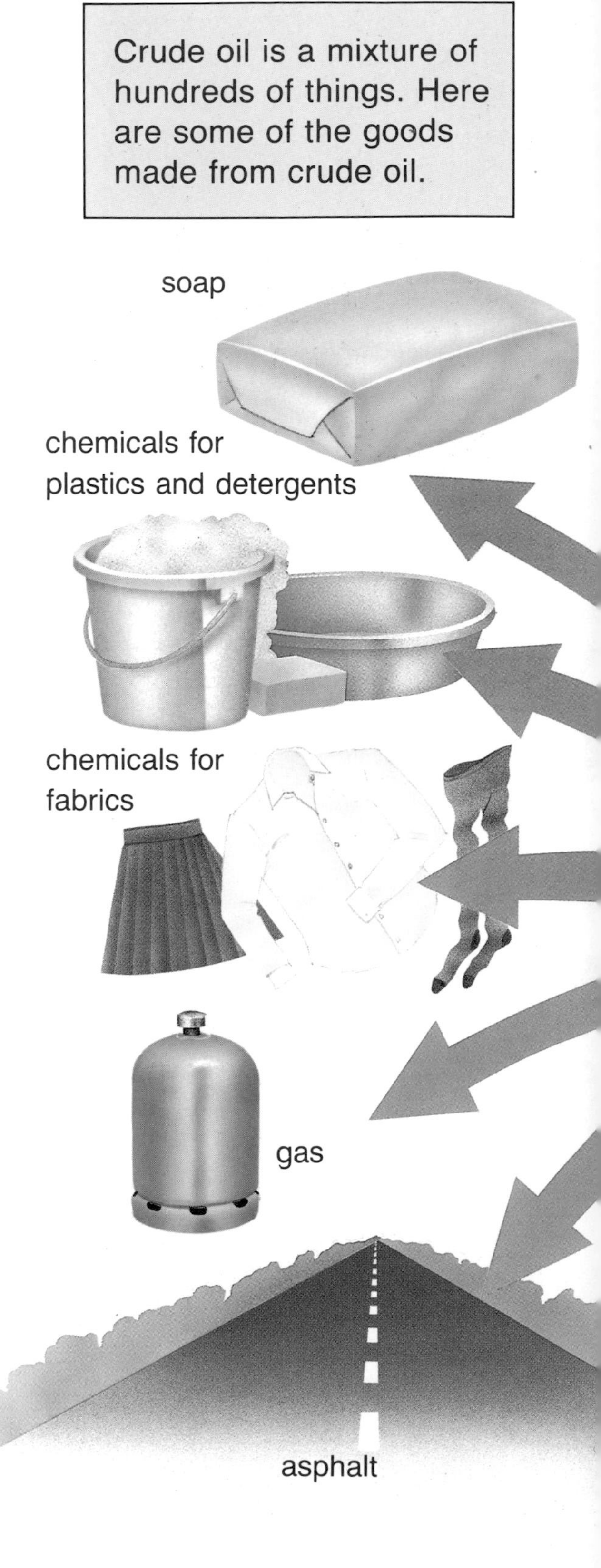

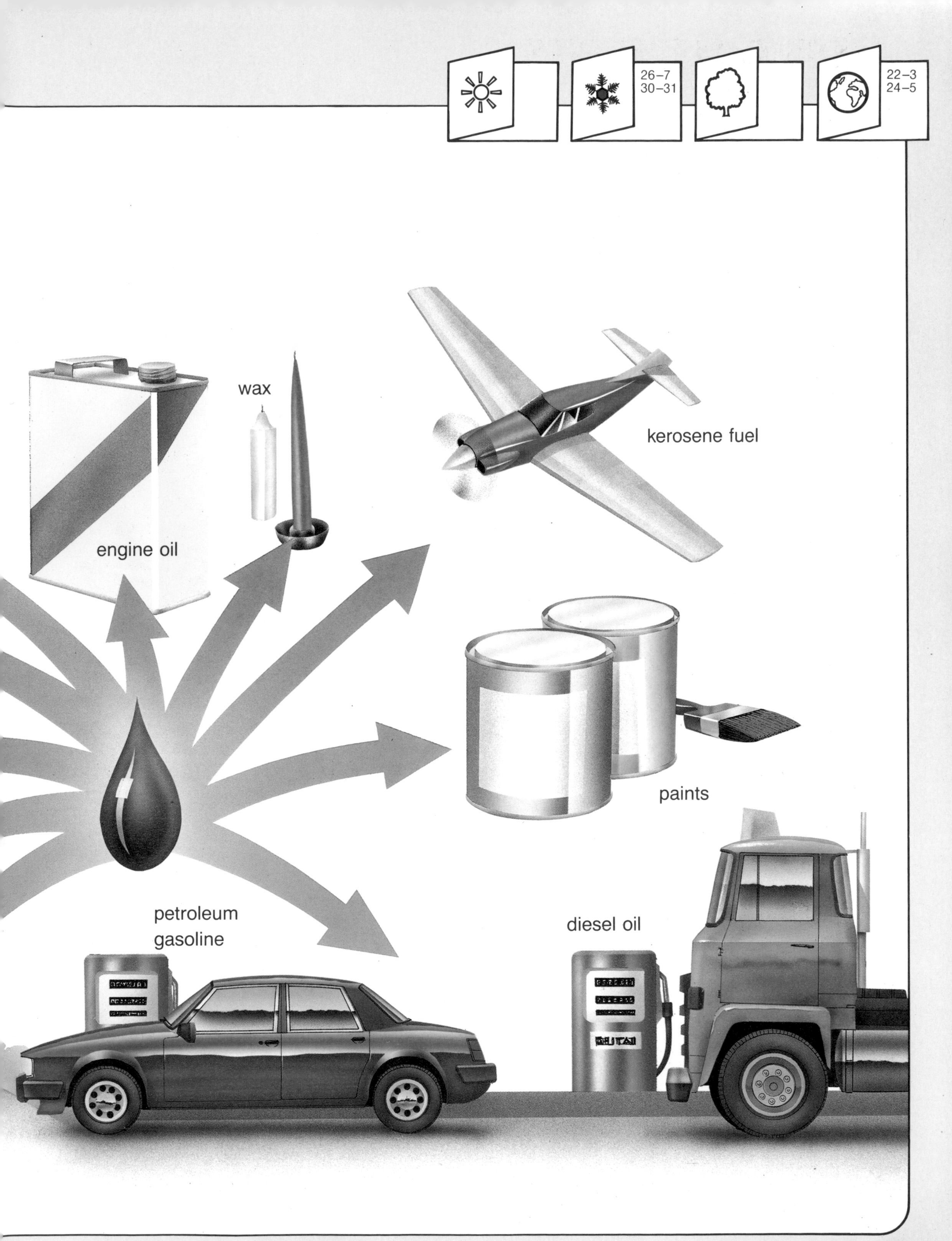
26–7
30–31
22–3
24–5
wax
kerosene fuel
engine oil
paints
petroleum
gasoline
diesel oil

How things are made

A long time ago, people made most things by hand. Now machines in factories do it instead. They can make things much faster than people can, and they never get tired like people.

Nearly all our machines are made of metal, and so are most of the things they make. Most metals are runny and red-hot when they come out of the furnaces where they are made. We can make this liquid metal into the shape we want by pouring it into a hollow mold, or cast. The metal cools and once it is solid, stays the same shape as the mold. We call this process **casting**.

All wool comes from sheep.

Washing the wool to get rid of dirt and grease.

Carding the wool takes out the tangles.

Dyeing the wool to color it.

Spinning the wool into fine threads.

We can also shape metal when it is solid. We can roll it with powerful machines, like mangles. We can hammer and squeeze it into shape with forges and presses. We can cut it with strong tools driven by electricity. We call this process **machining**.

Making things takes time. There may be several different stages to go through before you end up with what you want. The pictures show some of the different stages in making a woolen dress. It all began when the sheep grew the wool, and ended when the dressmaker sewed the woolen cloth together.

Shearing the sheep to get its fleece.

Weaving woolen cloth from thread.

Cutting out the cloth.

Dress sewn together from cloth.

The woolmark shows the dress is all wool. See if it is on any of your clothes.

Building

Can you imagine what it would be like with no houses? We would have to live outside, and we would get cold and wet and be very uncomfortable. This is why people began building homes for themselves thousands of years ago. They wanted somewhere to stay warm and safe from wild animals.

In the beginning, people built their homes from whatever they found around them, such as wood, stone, or mud. Some people still use natural materials like these. In Mexico and the Middle East, they build houses with bricks made of dried mud, or adobe. In Africa, people use grasses woven together to make them stronger.

Homes made from natural materials.

Stone and plaster house.

Mud-brick house.

Wooden stilt house.

Woolen cloth tent.

34–5
36
52
12–13
26–7

In most countries, people now make materials especially for building, such as bricks. We make bricks from clay and other things, and we bake them to make them hard. We stick the bricks together with mortar, which is a mixture of sand and cement. We use sand and cement to make **concrete**, too. Concrete is as hard as a rock once the mixture sets. We make office buildings and skyscrapers with it.

Concrete is also very heavy. It needs something to hold it up, or it will sag and break up. So we usually put steel rods into it to stop it from sagging. We also use steel to build the frame, or skeleton, of many buildings. It is so strong that it can hold up the tallest skyscraper.

Building a skyscraper in Tokyo. The skyscraper will be built around the frame of steel girders.

Animal builders

Animals build homes, too, just as people do. They are often very clever builders and are good at using whatever they find around them.

Most birds build their nests with natural materials, like twigs, moss, spiders' webs, and horses' hair. House martins and swallows make their nests of mud. They use straw and grass to help bind the mud together. Thrushes line their nests with mud to keep them dry and warm. Tailor birds actually sew leaves together to make a basket to hold the nest. But weaver birds are perhaps the best builders of all. They weave grasses together to make nests.

Weaver birds weave nests of grass. They even tie the grass in place.

Ants make different kinds of nests. Some stick leaves together with a sticky liquid which comes from their young, or larvae.

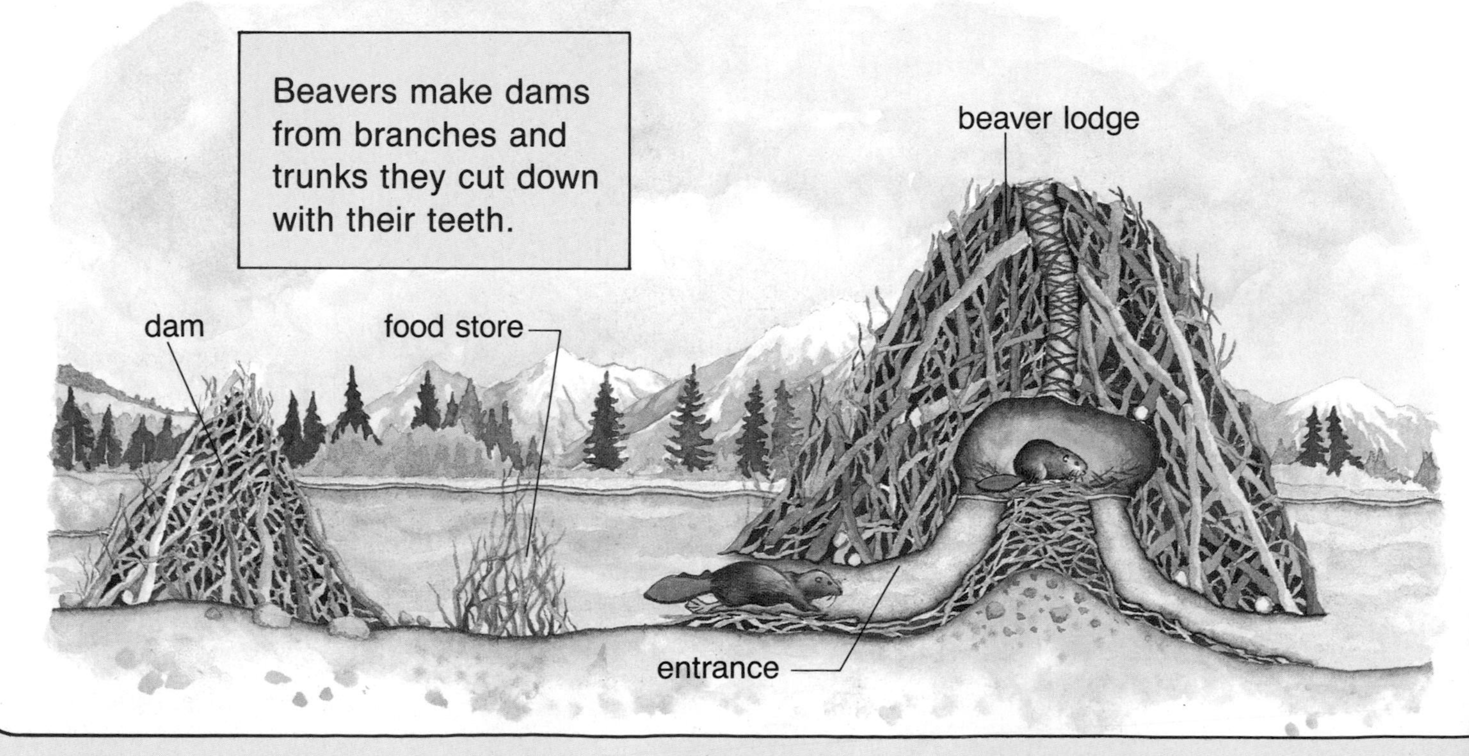

Beavers make dams from branches and trunks they cut down with their teeth.

Beavers make their homes with the entrance underwater because it is safer. So, to make sure the water is deep enough, they build **dams** across streams and lakes. This holds back the water and makes it deeper.

Some animals even make their own materials. Termites build huge mounds for themselves, several feet high. They cover the outside with hard stuff which they make from their droppings. It is almost as hard as concrete, and helps protect the nest.

Bees make wonderful nests. First, they make wax in their bodies. Then they use the wax to build hundreds of little boxes, or cells. This is the honeycomb that you get in some kinds of honey. The queen bee lays an egg in each of the cells.

There are many tunnels and rooms in termite mounds. Some are food stores, others are nurseries for the young.

6–7
32–3
34–5
8–9
24–5

Natural materials

We now make most of the materials we use for building. But one natural material, wood, is still very important to us. Wood is strong. The strongest parts of plants are the stems, trunks, and branches, and these are made of wood. Wood can make buildings strong, too.

In some ways, buildings are like trees. A tall tree has to be able to bend in the wind, or in the end it would break. A skyscraper also has to be able to move a little, although you do not notice it.

Trees have roots under them to hold them up and stop them from falling over or sinking into the ground. Buildings have layers of concrete, or **foundations**, underneath them to do the same thing.

This huge tree grows in the jungle. The roots grow out of the ground to give support to the trunk. You might see this sort of shape supporting the sides of some buildings.

ACTIVITY CONTENTS

Each activity has a list of things you will need to do it, so find them and have them all ready before you start.

Make sure you cover the table you're going to work on with old newspaper in case of any spills.

NEVER USE BOILING WATER. If you need to use hot water, get it from the tap.

NEVER TASTE ANYTHING you are using, unless the book tells you to.

Where the book tells you to get help from a grown-up then do so, because on your own the activity might be dangerous.

Growing crystals

You can easily grow crystals from things you can find at home. Start with table salt and then try other things such as alum and bath salts. You will have to use hot water some of the time, so be careful.

- A colored cup
- Colored saucers
- A small teaspoon
- A small cup
- Table salt, alum from the drug store
- Tweezers

1

Put a small cupful of water from the hot tap into the cup. Add a level teaspoonful of salt and stir to help the salt dissolve.

2

Add more salt until no more will dissolve and some is left in the bottom of the cup. You have made a salt solution. How much salt have you added?

3

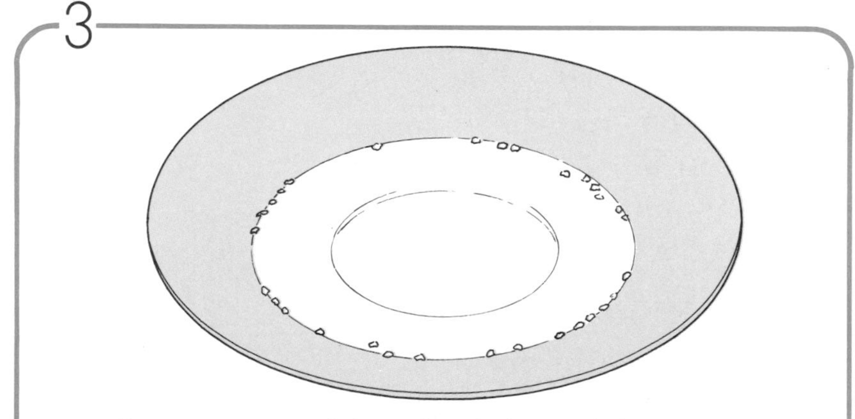

Pour out some of the salt solution so it just covers the bottom of a saucer. Put the saucer in a warm place. Look at the saucer later on and the next day, to see what is happening.

4

Soon you will see crystals of salt on the bottom of the saucer. Later, you will see that the water is slowly disappearing, or evaporating into the air, and there are more salt crystals. Look at them carefully. What shape are they?

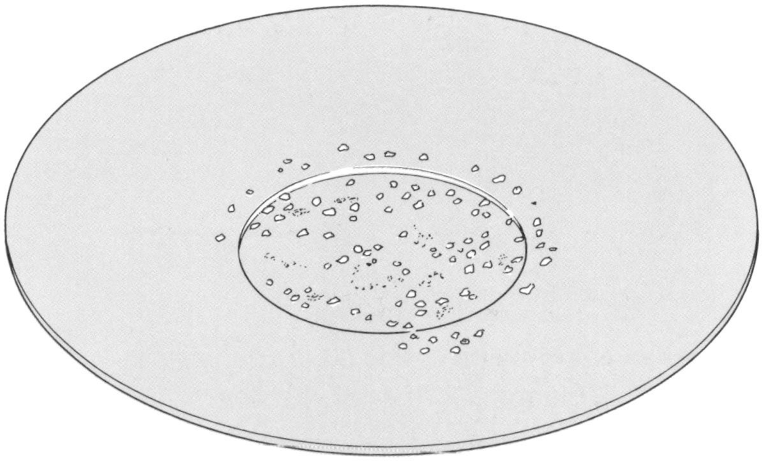

Try this again but dissolve the salt in cold water instead of hot. How many spoonfuls of salt can you dissolve in the cold water? Does it take more or less time for the crystals to grow in the saucer this time?

Try the experiment using alum, instead of salt, dissolved in a small cupful of hot water. When you see crystals starting to grow in the saucer, pour the liquid off into a clean saucer. Then carefully use the tweezers to pick out the 2 or 3 biggest crystals left in the first saucer, and put them into the solution in the new saucer. Leave the saucer in a warm place again and see how big the crystals can grow. What shape are the alum crystals? You could keep a crystal in a scrapbook with a piece of tape.

Gases all around?

When you pick up an empty bottle, are you sure it's empty?

- An empty bottle
- A bucket of water

1 Hold an empty bottle upside-down and push it slowly into the water. What can you see?

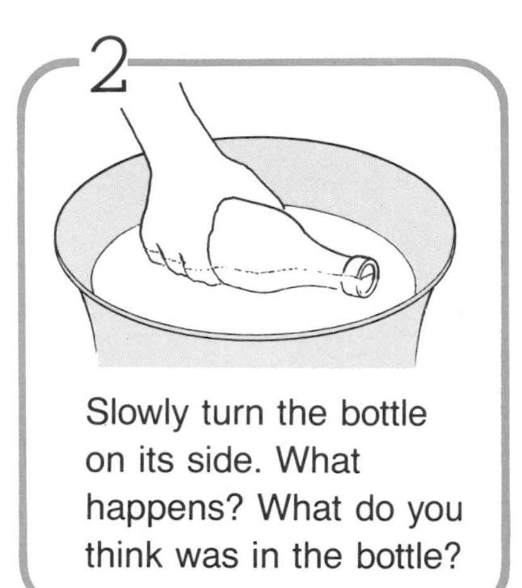

2 Slowly turn the bottle on its side. What happens? What do you think was in the bottle?

Can air slow you down? Try this out on a day when there isn't much wind.

- A watch or clock with second hand
- A piece of stiff cardboard or a light, plastic tray about twice as wide as your body

1 Mark out a distance of 100 strides to run. Get a friend to time how long it takes you to run it.

2 Now run the same distance again but holding the cardboard in front of you. Did it take you any longer to run? Did it feel as if something was pushing against you?

What do you think happened. Did air slow you down? Can you think of what this might do to cars, buses, and trucks?

Water on the level

12–13

What happens to liquid in a container when you move it around? You can try it with water here, but all liquids do the same.

- An empty glass
- A see-through plastic tube, about 5 feet long

1

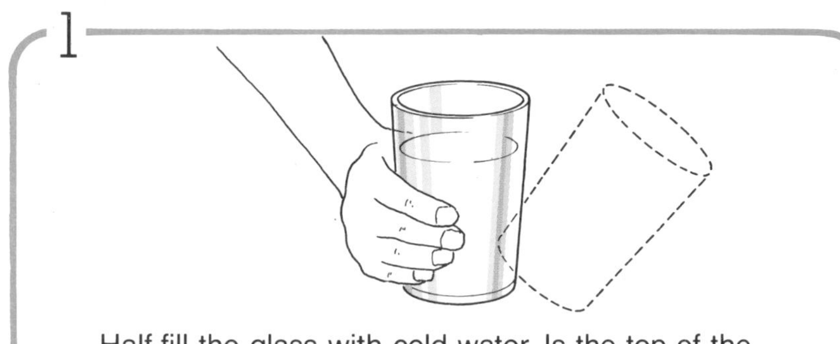

Half-fill the glass with cold water. Is the top of the water flat? Tip the glass to one side. What do you think will happen to the water? Is the top flat now?

2

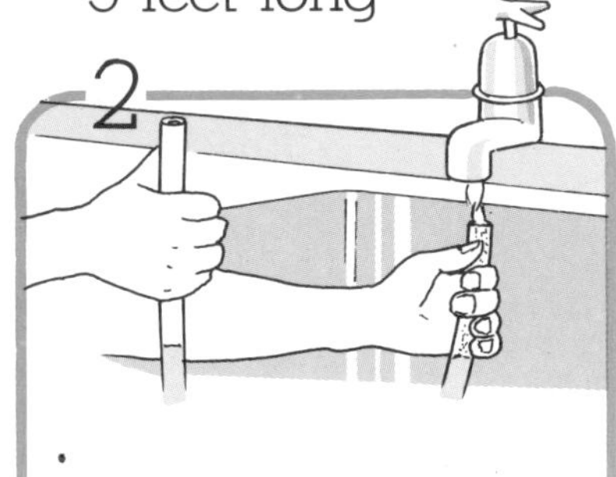

Fill the plastic tube with water from the cold tap. Keep the tube in the sink and hold the other end up.

3

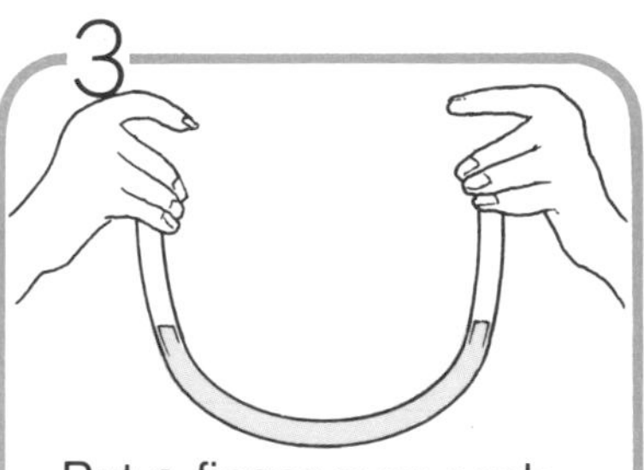

Put a finger over each end of the tube to close it. Hold it up in a U-shape, with the two ends of the tube at the same height. Is the top of the water in one side as high as it is in the other side? Is the top flat?

4

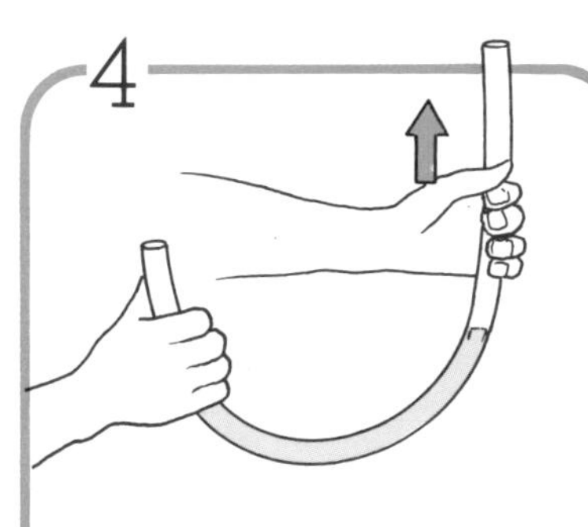

Move one end of the tube up. What happens to the top of the water in each side? Now move one end of the tube down and watch the water again.

5

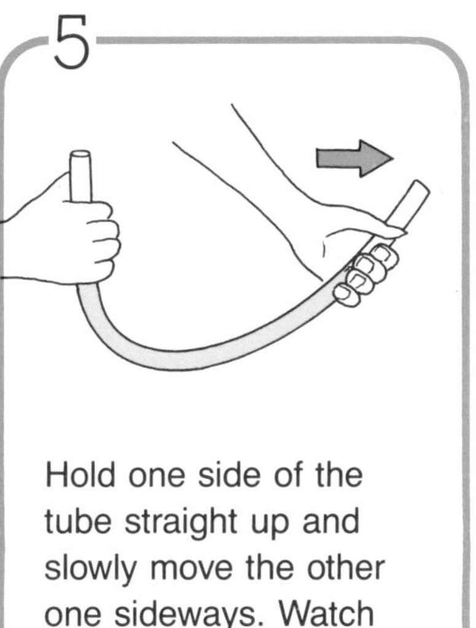

Hold one side of the tube straight up and slowly move the other one sideways. Watch what happens to the water. Is the top flat or sloping?

Did the water do what you expected? What do you think would happen if you had something solid, such as stones, in the glass or tube when you tilted it?

Water pressure

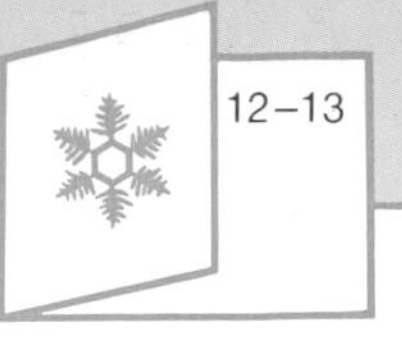
12–13

Does deep water push harder than shallow water?

- An empty large size plastic soft-drink bottle, or liquid detergent bottle
- A sharp nail

1

Fill the bottle with cold water from the tap. Keep the bottle in the sink all the time, with water trickling into the bottle from the tap. With a sharp nail, make a hole in the side of the bottle about half-way down. (You might need help from a grown-up to make the hole). What happens?

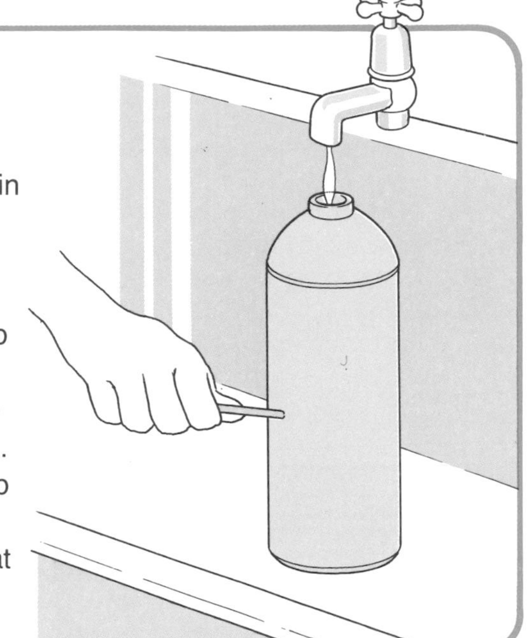

2

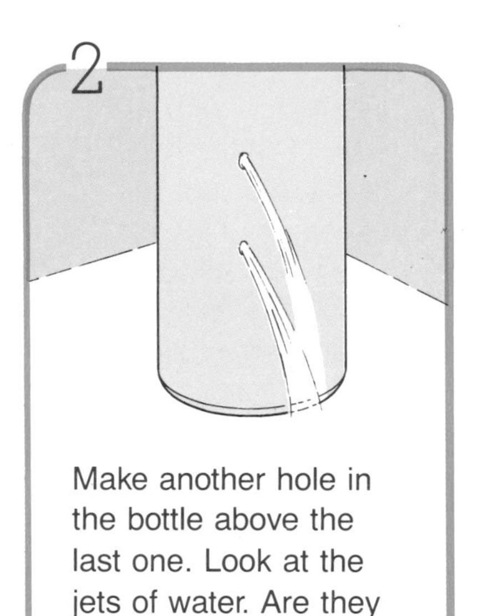

Make another hole in the bottle above the last one. Look at the jets of water. Are they the same or different?

3

Keep the tap running so the bottle is still full. Make another hole in the bottle, lower than the other holes, and make one more near the bottom of the bottle. What are the water jets like now? Put your hand in the jets and see if any seem to push harder.

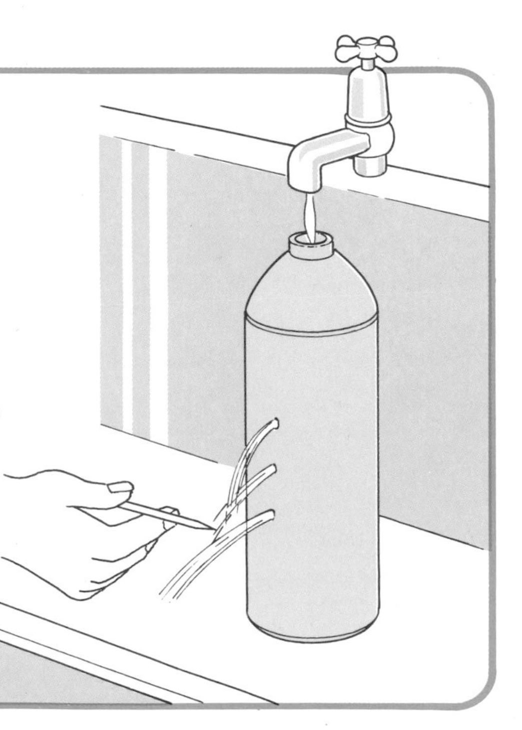

What do you think this shows about the push of the water, or water pressure? How do you think it would feel to be deep down in the sea?

Float a needle

Everyone knows that steel is heavier than water. So, if you put a steel needle in water it will sink. Or will it?

- A piece of tissue paper
- A needle
- A shallow dish or saucer

1

Fill the dish with water. Put the needle on the piece of tissue paper, and float the paper on the water in the dish. What happens to the paper and the needle? Look closely at the needle and the water surface. What can you see? What do you think has happened?

2

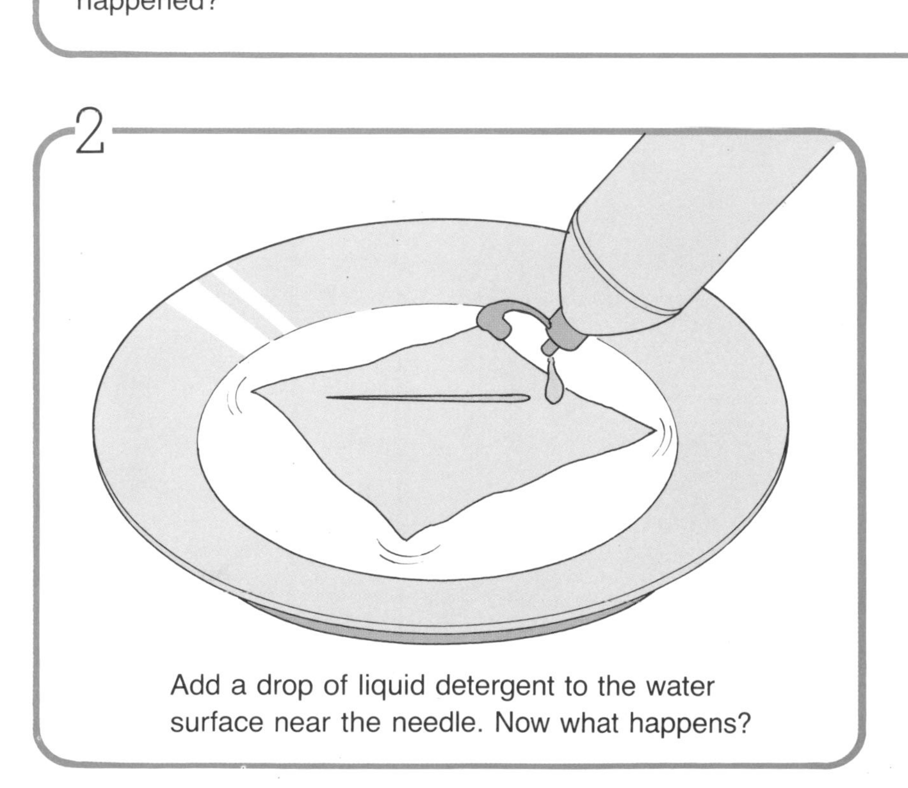

Add a drop of liquid detergent to the water surface near the needle. Now what happens?

What makes things float?

Why do some things float and others sink?
How can a heavy steel ship float?

- An empty drink can
- Modeling clay
- Coins or marbles the same size

1

Collect as many things as you can which are about the same size and shape. Find a piece of wood, a stone, a piece of brick, a metal weight for the scales, and a piece of styrofoam. Try to find things without air inside them. How does each one feel in your hand?

2

Fill the sink with water. Test the things you collected and write down which float and which sink. Do you think that floating depends on what a thing is made of, what shape it is, or what size it is? Do all pieces of wood float? Do all metal things sink?

3

Now see if the empty drink can floats. What happens if you squash the can flat and then put it in the water?

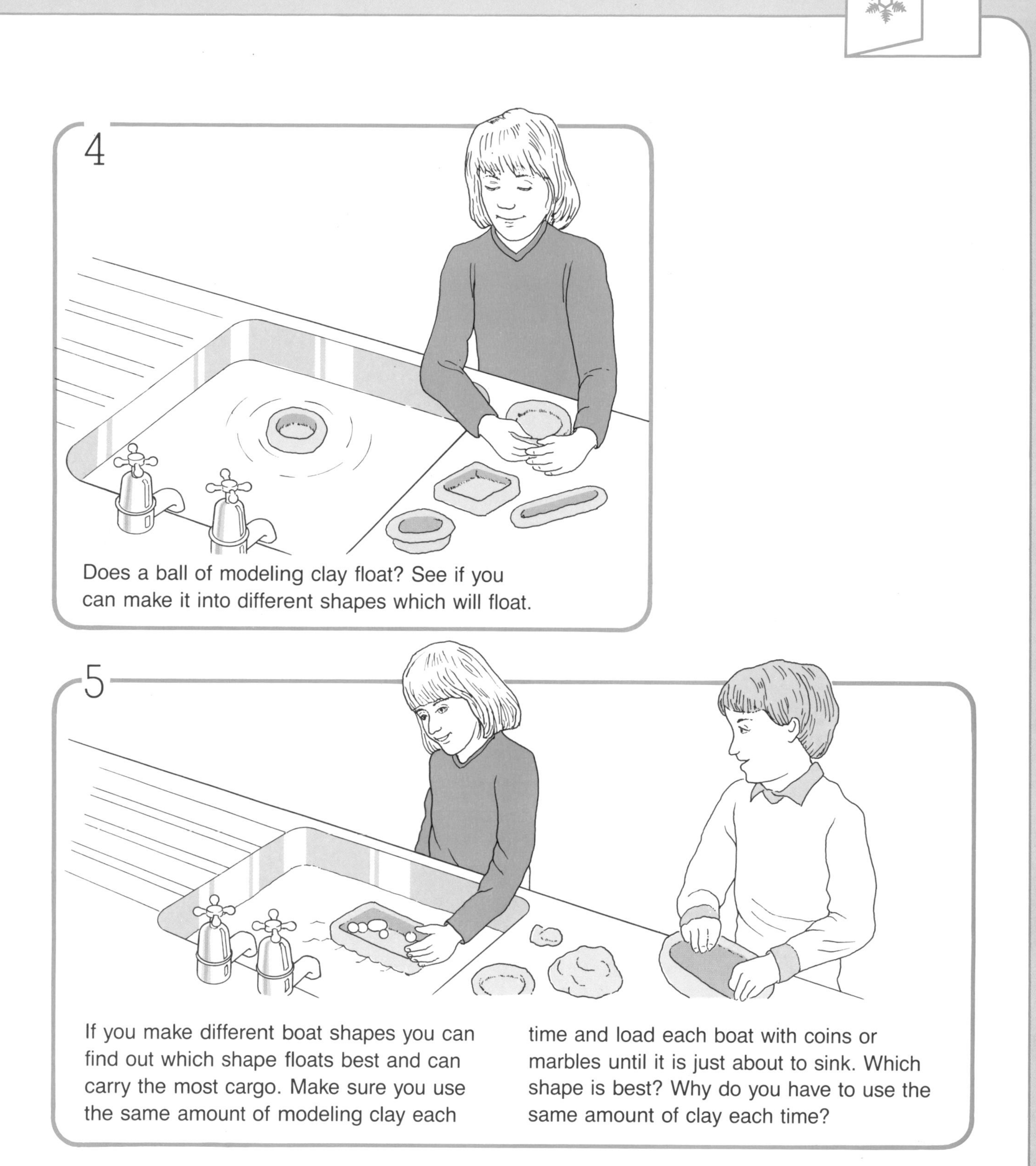

4

Does a ball of modeling clay float? See if you can make it into different shapes which will float.

5

If you make different boat shapes you can find out which shape floats best and can carry the most cargo. Make sure you use the same amount of modeling clay each time and load each boat with coins or marbles until it is just about to sink. Which shape is best? Why do you have to use the same amount of clay each time?

A lump of steel sinks, but, like the modeling clay and the metal drink can, it can be made into a shape which will float — like a big ship.

Separating sand from salt

22–3

During the winter, trucks spread a mixture of salt and sand on icy roads. This mixture helps melt the ice and gives a better grip for tires. The salt used for the roads is a dirty colored rock salt, which is mined from the ground. Suppose someone gave you a dirty sand and salt mixture like this and asked you to get pure salt from it. What would you do?

- A mixture of salt and sand
- A kitchen funnel, or the top cut off a plastic soft-drink bottle
- Filter paper, for a filter coffee maker or blotting paper
- Old mugs or cups
- A tablespoon
- A saucer

1

Put 4 or 5 spoonfuls of the sand and salt mixture into the mug and add the same amount of hot water. Stir well. When it stops swirling around what can you see? What do you think has happened to the salt?

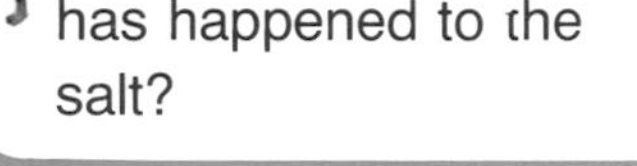

2

Put the funnel in another mug. Fold a filter paper so that it fits into the funnel.

3

Carefully pour the mixture from the first mug into the filter paper in the funnel. Is the liquid that comes out of the funnel clean? What is left on the filter paper?

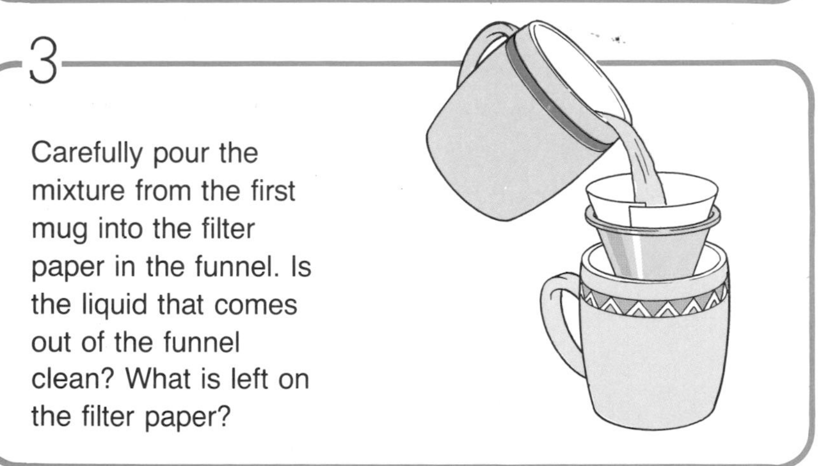

4

Pour out some of the solution in the second mug into a saucer. Leave it in a warm place and see what happens. Taste a tiny bit of what is in the saucer. What is it?

What have you gotten from the sand and salt mixture you started with? Have you ever seen salty white marks left on cars after water splashes up from the roads?

Separating the colors

22–3

What colors make up colored inks? Is the ink in a black felt-tip pen really black?

- Pieces of white blotting paper, or filter paper from a coffee machine
- Different colored felt-tip pens
- Food colorings
- A jelly jar

1

Cut a square of blotting paper with sides about 4 inches long. On one side make two cuts, about ½ inch apart, into the middle.

2

Use a felt-tip pen to make a black spot about ½ inch across in the middle, at the top of the two cuts. Bend down the strip of paper made by the cuts. Then put some water in the jar and rest the blotting paper on top of the jar. Make sure the strip of paper hanging down just dips into the water. Watch the water spreading up the strip and across the blotting paper. What happens? Can you see more than one color?

Try this again on other pieces of blotting paper with different felt-tips, or with a drop of food coloring. Try mixing two food colorings together. Can you get the separate colors back again?

When the blotting paper has dried, you could stick it in a scrapbook. Write down what color you first put on the paper.

To rust or not to rust

Rusting is one of the chemical changes that take place around us all the time. What makes things rust?

- Some small jars or yogurt cups
- Some steel nails
- Salt
- Vinegar
- Sandpaper or steel wool
- Grease such as Vaseline or cold cream
- Clear nail polish
- A galvanized nail — one covered in zinc

1

Pour enough tap water into three of the jars to cover a nail. Put a teaspoon of salt into one and mix it in. Add about 2 teaspoonfuls of vinegar to another. Mark the jars to show what each holds.

2

Clean four nails with sandpaper until they are very shiny. Put one into each of the three jars. Put the fourth into an empty jar. Look at the jars every day for a week, and write down what happens to the four nails and when. Do they all get rusty?

Does any nail get rusty faster than another?

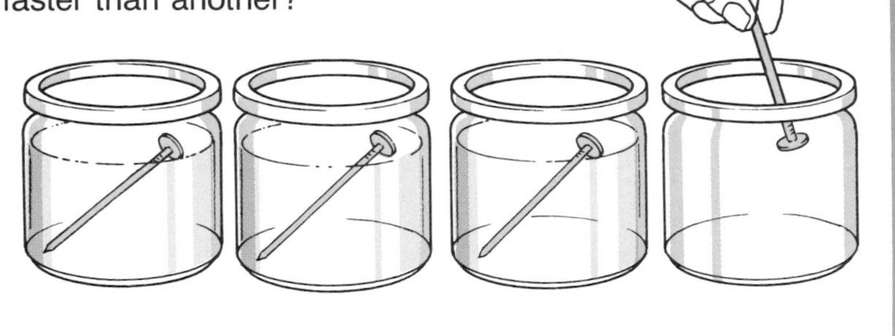

Try this again. Use the mixture which made the nail get rusty fastest. Use 3 jars and leave one in the same place as you did before. Put another jar in the refrigerator and leave the third in a warm place, such as by a radiator. Where does rusting happen most quickly and where most slowly? Do you remember from page 46 about the sand and salt mixture for roads? So why should drivers wash the underneath of their cars after winter is over?

How can you stop things from getting rusty?

Leave the nails in the salt water as long as you did the first time. Have the nails gotten rusty? Why do you think painting a nail or covering it with grease or another metal stops rusting?

Sugar and yeast

Our bread is light and airy because the yeast in it makes some chemical changes. Yeast is a living thing that feeds on sugar. We can find out something about how yeast works.

- Dried yeast – use the sort for baking
- A pitcher
- A teaspoon
- Sugar
- Two balloons
- Two clean bottles with narrow necks

1

Half fill the pitcher with warm (not hot) water from the tap, and add four teaspoonfuls of sugar. Stir until the sugar dissolves. Dip your finger into the solution and see how it tastes. Pour the sugar solution into the bottle. Label it.

2

Add a heaped teaspoonful of dried yeast to the solution in the bottle. Shake the bottle.

3

Put a teaspoon of yeast into another bottle. Half fill the bottle with warm water and shake it.

What happens to the balloons? Look for bubbles in the jars. Are both jars the same, or has the sugar made a difference? What is the yeast doing?

Leave the bottle with bubbles for the yeast to carry on working. Eventually it will stop working, and no more bubbles will be given off. Now put your finger in the solution and taste it. Is is still sweet? Has the yeast used the sugar as it made the bubbles? Why do we use yeast when we make bread?

Does the temperature make any difference to the yeast? Make up the sugar solution and put it in the refrigerator to get very cold, then add the yeast. Keep the bottle in the refrigerator. Also make up the sugar solution with very hot tap water and then add the yeast to the hot water. What happens this time? Yeast doesn't work well in the cold, and hot water kills it.

The importance of shape

32–3

When we build things, we must use the right kind of material to make them strong. We must also use that material in the right way. Find out which shapes are strongest.

- Sheets of stiff paper
- Cardboard (cut from a cereal box)
- A matchbox
- A piece of string
- Several coins the same, to use as weights
- Some books

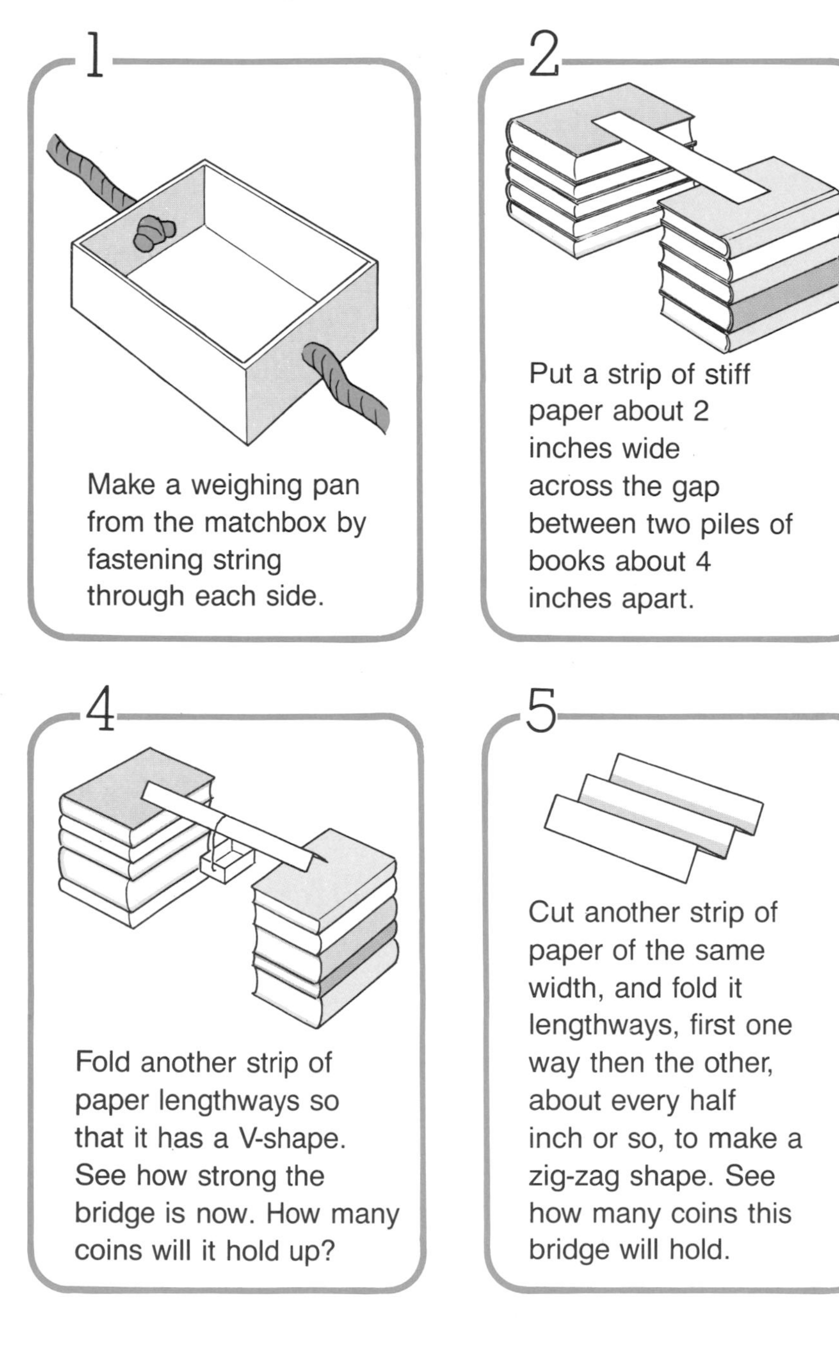

1

Make a weighing pan from the matchbox by fastening string through each side.

2

Put a strip of stiff paper about 2 inches wide across the gap between two piles of books about 4 inches apart.

3

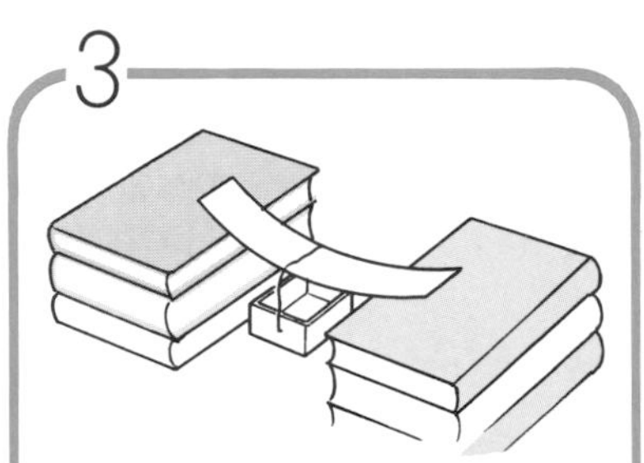

Hang the box from the paper bridge. How many coins can you put in before the bridge falls down? Write it down.

4

Fold another strip of paper lengthways so that it has a V-shape. See how strong the bridge is now. How many coins will it hold up?

5

Cut another strip of paper of the same width, and fold it lengthways, first one way then the other, about every half inch or so, to make a zig-zag shape. See how many coins this bridge will hold.

6

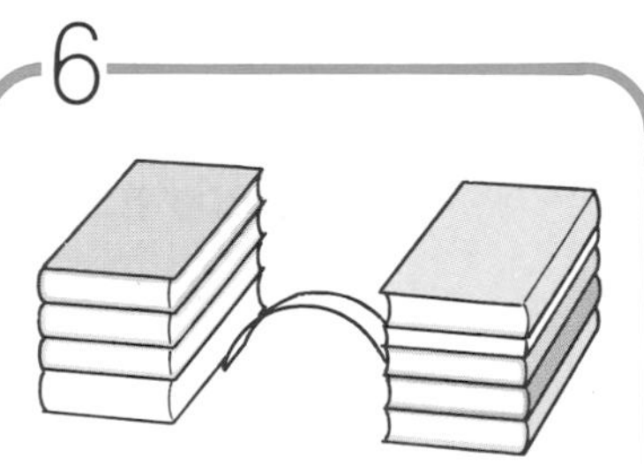

Try all these tests again using strips of cardboard. See what weights the different shapes can hold. Make an arch shape with cardboard by fixing it between the piles of books. See how strong that is.

Which was stronger — paper or cardboard?
Which was the strongest shape?

GLOSSARY

A glossary is a word list. This one explains the unusual words that are used in this book.

Alloy A mixture of two or more metals. Steel and bronze are very useful alloys.
Bronze An alloy which is a mixture of copper and tin.
Casting A way of shaping red-hot liquid metal inside a mold.
Compressed air When you pump up a bike tire, you force lots of air inside. The air inside becomes squeezed, or compressed.
Concrete One of our main building materials. We make it by mixing together cement, sand, and gravel with water. The mixture becomes rock-hard when it sets.
Corrode To eat away. Acids and salty water corrode metals, making them rusty and pitted. Acids cause corrosion.
Crystal A shiny, glassy material. Minerals form crystals in rocks. Water makes crystals of ice when it freezes.
Dam A barrier built across a river. It holds back the water.
Dissolve Sugar dissolves when you stir it into water. It disappears into the water, forming a solution.
Dry cleaner A liquid that dissolves grease. It doesn't contain any water.
Foundations Workers build foundations in the ground beneath houses and other structures. The foundations stop the buildings from sinking and toppling over.
Hard water Water with lots of minerals dissolved in it.
Machining Cutting metals to shape with power-driven tools, called machine tools.
Minerals Materials that make up the rocks. We can dig minerals from the ground in mines.
Natural gas Gas that we find trapped in the rocks. We use it as a fuel.
Petroleum The oil we find trapped in rocks. Its name means "rock oil." It is our most important fuel and raw material.
Plastic A material that we can easily shape by heating. Most plastics are made from oil.
Raw material Something that is of little use by itself but we can turn it into other useful things.
Refining Making purer. Petroleum is refined into hundreds of useful products at an oil refinery.
Reservoir An artificial lake — one that didn't form naturally. A reservoir forms behind a dam, and stores water.
Rust The powdery, orange-brown coating that forms on iron in dampness. It means the iron is corroding.
Skyscraper A building so high that it seems to touch the sky.
Soft water Water that doesn't have many minerals dissolved in it.
Solution A liquid which has something dissolved in it.
Steel An alloy that contains iron and traces of other metals. It is probably our most important metal.
Surface tension The skin on the surface of liquids.

INDEX

The **dark** numbers tell you where you will find a picture of the subject.